Everyday Science Experiments in the Car

John Daniel Hartzog

The Rosen Publishing Group's
PowerKids Press™
New York

Some of the experiments in this book are designed for a child to do together with an adult.

Published in 2000 by The Rosen Publishing Group, Inc.
29 East 21st Street, New York, NY 10010

Photo Credits and Photo Illustrations: pp. 4, 6, 7, 8, 10, 12, 13, 19 by Shalhevet Moshe; p. 11 © Peter Langone/International Stock; p. 15 © Roger Markham-Smith/International Stock; p. 16 © Scott Campbell/International Stock; p. 20 © Maya Krome/International Stock.

First Edition

Book Design: Michael de Guzman

Hartzog, John Daniel
      Everyday science experiments in the car / by John Daniel Hartzog.
         p.   cm. — (Science surprises)
      Includes index.
      Summary: Provides simple experiments that explore scientific phenomena occuring in a moving car.
      ISBN 0-8239-5459-5 (alk. paper)
      1. Science—Experiments—Juvenile literature. [1. Science—Experiments. 2. Experiments.] I. Title. II. Series: Hartzog, John Daniel. Science surprises.
Q164.H274   1999
507'.8—dc21   99-13879
                                                                       CIP

Manufactured in the United States of America

Contents

What Is Car Science?

People spend lots of time in their cars. We know our cars pretty well, right? There is actually a lot you can learn about your car, though. For example, you probably don't know how your car runs, or why drops of water make lines on car windows. Scientists discover new things about **familiar** objects every day. Science helps us learn new things and understand our world by helping us make discoveries. Discoveries are what happen when someone sees, hears, or thinks about something in a new way. These **experiments** will help you to make discoveries in your car.

*Remember to wear your seatbelt while you complete the experiments in this book.

◄ Most families spend a lot of time in their cars.

Sense of Direction

Have you ever noticed that when you are in your car, you can tell what direction you are moving in, even with your eyes closed? That is because your body has a special **organ** inside your ear to help you sense direction and movement. This organ is called your inner ear. It is full of a liquid that is like water. When the liquid moves, **nerves** tell your brain which direction you're going in.

Let's try an experiment to help you see how your inner ear works. On your next car trip, hold a half-full bottle of water straight up. As the car

Materials Needed:
• One half-full bottle of water

moves, watch the liquid move inside the bottle. When your car turns a corner, the liquid moves to that side. If the car stops quickly, the water moves forward. Now put the bottle down and close your eyes. Can you feel your inner ear telling you which way you are going? Tell the driver what you think and see if you got it right!

Does the water in the bottle tilt as you move? ▶

- One small ball

8

Inertia

When a car is moving, it will keep moving even if the driver takes her foot off the gas pedal. This is because of **inertia**. Inertia is a **force**. It keeps still things where they are and keeps moving things moving. You can feel this force by leaning forward slowly as the car begins to go faster. The inertia will try to keep you still by holding you in your seat. Place a ball on the floor of the car. When the car begins to move, the ball rolls backwards. The ball is moving within the car, but inertia is trying to keep it over the same piece of ground as it was when the car was stopped. When the car stops moving, the ball rolls forward. The inertia makes the ball keep moving even though the car isn't.

◀ *Place the ball on the floor of the car and watch how it moves.*

9

Car Classification

There are hundreds of different kinds of cars made every year. Telling the difference between cars can be difficult. Scientists use **classification**

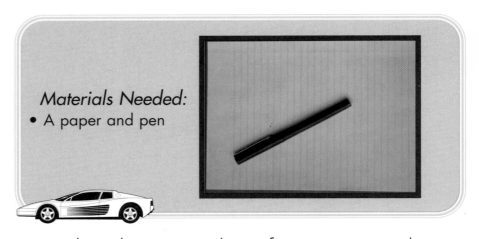

Materials Needed:
• A paper and pen

to tell the difference between similar things. Classification is when you place things in groups based on similarities they have with each other. We can use classification to see the differences between cars.

Let's make up a system of classification for the cars you see. First make a list of things that all cars have, like four wheels or headlights. Then write a list of differences, like the number of

This car has only two doors. ▶

doors, the height, or the shape of cars. Can you tell the difference between sports cars and family cars? Sports cars usually have only two doors, and they are lower to the ground. Family cars can be used to carry lots of people. Try to write a list of different classes of cars, such as sports cars, family cars, four-door cars, and two-door cars, and count how many you see of each. Classification makes it easy to identify the different kinds of cars.

◀ *Would you classify this car as a sports car or a family car? Why?*

The Theory of Relativity

Another scientific idea you can study in the car is Albert Einstein's **Theory** of Relativity. In science, relativity means that each person's observations will be different depending on how each person sees something. Let's try an example in the car.

Materials Needed:
- Something you can throw (a bean bag or ball)

Bring something with you into the car that you can throw gently into the air and catch. If you toss it as the car is moving, you will see it travel up and then back down again. If someone standing by the side of the road watched you toss the object, though,

12

he would see it go up and down, but he would also see it traveling forward, since the car is moving forward. Each observation is relative to the observer's view of the experiment.

To you, the ball seems only to be going up and down, but to someone standing by the side of the road, it would seem to be moving forward as well.

The Doppler Effect

Have you ever noticed how the sound of a siren or horn changes as it moves past you? Next time a police car or ambulance passes you with its sirens on, listen closely. As the noise gets closer, it sounds higher, like a scream. As the noise gets farther away, it gets lower, like a deep moan. This is because sound travels in **waves**. If you could see them, these waves would look a lot like the waves on a lake or in the ocean. Sound waves, though, can get pushed closer together or stretched farther apart. This pulling and pushing is called the **Doppler effect**. If you are in a car, you can often hear its effect when a siren's sound changes.

Listen closely next time an ambulance or police car drives past you. ▶

Car engines run on little explosions of gasoline.

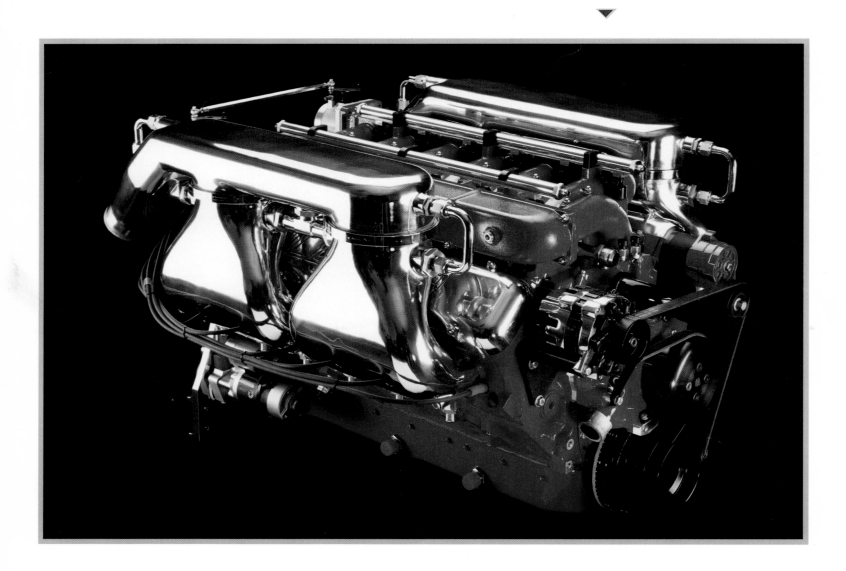

The Car Engine

You probably know that cars run on gasoline, but have you ever wondered how gas makes your car go? Gasoline is a **fuel**. A fuel is something that burns easily. The car's engine burns the gasoline to move the car. Here's how it works: When someone steps on the gas pedal, the gas enters the engine. The pressure of the gas moving into the engine causes a little spark. Since gasoline burns easily, the spark causes a small explosion. The pressure from the explosion pushes a gear, which turns a metal rod called an **axle**. The axle connects the wheels, so when the axle turns, the wheels do, too. That's how gas makes your car go!

Perspective

When you are riding in the car, do you ever notice that a building that is far away looks very small? As you get closer, though, the building seems to get bigger and bigger. That's because your **perspective** changes as you move. Perspective is how distance affects the things you see. Next time you are in the car try this experiment. Pick a building, a tree, or some large object in the distance. Hold your arm straight out in front of you. Close one eye and compare the size of the object to the length of your finger. Does the object look smaller than your finger? Watch the object closely as you drive towards it. Does it start to look larger than your finger? Your perspective of the building changed as you got closer to it.

Compare the size of your finger with the size of the object in the distance.

As you get closer, the object will go from being smaller than your finger to being larger than it.

*The patterns water drops make on car windows
help car designers build better cars.*

▼

Water Movement

Do you ever notice how drops of water move on the car windows when you drive in the rain? If you watch them closely you will begin to see a **pattern**. The path these droplets take depends on the air that moves around the car as it drives. While you are moving forward, the drops move toward the back of the car. When the car is stopped, the drops will go straight down, or in the direction the wind is blowing. Try rolling your window down just a little as the car is moving. This changes how the air moves past the window, since some can get into the car. Do the drops move differently with the window open? Car designers use experiments with wind and water movement to help them find the best shapes for cars. If the water moves back in straight lines, the designers know the car is fast and well-designed.

Science in the Car

Cars wouldn't exist today if scientists hadn't figured out how to make them run. Cars move faster and differently than we can on our own. They open up a whole world of mysterious things to discover, like inertia, the Doppler effect, and the Theory of Relativity. By using experimentation, classification, and observation, science allows us to explore the world of the car and the world we live in every day.

Glossary

axle (AK-sul) The metal rod that connects the wheels of the car.

classification (KLAS-ih-fih-kay-shun) The arrangement of things based on shared traits.

Doppler effect (DOP-luhr uh-FECT) A change in the frequency of a sound when the listener and the sound move toward or away from one another. Noises sound higher in frequency when traveling toward the listener and lower when moving away from the listener.

experiments (ek-SPER-ih-ments) Scientific tests designed to answer questions.

familiar (fuh-MIL-yer) Well-known.

force (FORS) The power or energy used to do something.

fuel (FYOO-el) Something that is used to provide another thing with energy.

inertia (in-UR-shuh) The tendency of an object at rest to stay at rest or an object in motion to stay in motion.

nerves (NERVZ) Thread-like fibers that carry messages through your body from sensory organs like your skin and eyes, to your brain, and from your brain to your muscles. Nerves help you to feel, move, and think.

organ (OR-gen) A part of a plant or animal designed for a certain function, like the heart.

pattern (PA-turn) A way things happen that is repeated in the same way again and again.

perspective (pur-SPEK-tihv) The effect of distance on the appearance of objects.

theory (THEE-ree) An explanation of how or why something happens.

waves (WAYVZ) The up-and-down movement of something through a substance such as air or water.

Index

Web Sites:

You can learn more about science experiments on the Internet. check out this Web site:
http://www.vickicobb.com